SIDELIGHTS ON RELATIVITY

Albert Einstein

Dover Publications, Inc.
New York

This Dover edition, first published in 1983, is an unabridged and unaltered republication of the translation by G. B. Jeffery, D.Sc., and W. Perrett, Ph.D., first published by E. P. Dutton and Company, Publishers, N.Y., in 1922.

Manufactured in the United States of America
Dover Publications, Inc., 180 Varick Street, New York, N.Y. 10014

Library of Congress Cataloging in Publication Data

Einstein, Albert, 1879–1955.
Sidelights on relativity.

Reprint. Previously published: New York : E. P. Dutton, 1923.
1. Relativity (Physics)—Addresses, essays, lectures.
I. Title.
QC173.58.E55 1983 530.1'1 83-5275
ISBN 0-486-24511-X

Contents

1
ETHER AND THE THEORY OF RELATIVITY
An Address delivered on May 5th, 1920, in the University of Leyden

25
GEOMETRY AND EXPERIENCE
An expanded form of an Address to the Prussian Academy of Sciences in Berlin on January 27th, 1921.

ETHER AND THE THEORY OF RELATIVITY

An Address delivered on May 5th, 1920,
in the University of Leyden

ETHER AND THE THEORY OF RELATIVITY

HOW does it come about that alongside of the idea of ponderable matter, which is derived by abstraction from everyday life, the physicists set the idea of the existence of another kind of matter, the ether? The explanation is probably to be sought in those phenomena which have given rise to the theory of action at a distance, and in the properties of light which have led to the undulatory theory. Let us devote a little while to the consideration of these two subjects.

Outside of physics we know nothing of action at a distance. When we try to connect cause and effect in the experiences which natural objects afford us, it seems at first as if there were no other mutual actions than those of immediate contact, e.g. the communication of motion by impact, push and pull, heating or inducing combustion by means of a flame, etc. It is true that even in everyday experience weight, which is in a sense action at a distance, plays a very important part. But since in daily experience the weight of bodies meets us as something constant, something not linked to any cause which is variable in time or place, we do not in everyday life speculate as to the cause of gravity, and therefore do not become conscious of its character as action at a distance. It was Newton's theory of gravitation that first assigned a cause for gravity by interpreting it as action at a distance, proceeding from masses. Newton's theory is probably the greatest stride ever made in the effort towards the causal nexus of natural phenomena. And yet this theory evoked a lively sense of discomfort among Newton's contemporaries, because it seemed to be in conflict with the principle springing from the rest of experience, that there can be reciprocal action only through contact, and not through immediate action at a distance.

It is only with reluctance that man's desire for knowledge endures a dualism of this kind. How was unity to be preserved in his comprehension of the forces of nature? Either by trying to look upon contact forces as being themselves distant forces which admittedly are observable only at a very small distance—and this was the road which Newton's followers, who were entirely under the spell of his doctrine, mostly preferred to take; or by assuming that the Newtonian action at a distance is only *apparently* immediate action at a distance, but in truth is conveyed by a medium permeating space, whether by movements or by elastic deformation of this medium. Thus the endeavour toward a unified view of the nature of forces leads to the hypothesis of an ether. This hypothesis, to be sure, did not at first bring with it any advance in the theory of gravitation or in physics generally, so that it became customary to treat Newton's law of force as an axiom not further reducible. But the ether hypothesis was bound always to play some part in physical science, even if at first only a latent part.

When in the first half of the nineteenth century the far-reaching similarity was revealed which subsists between the properties of light and those of elastic waves in ponderable bodies, the ether hypothesis found fresh support. It appeared beyond question that light must be interpreted as a vibratory process in an elastic, inert medium filling up universal space. It also seemed to be a necessary consequence of the fact that light is capable of polarisation that this medium, the ether, must be of the nature of a solid body, because transverse waves are not possible in a fluid, but only in a solid. Thus the physicists were bound to arrive at the theory of the "quasi-rigid" luminiferous ether, the parts of which can carry out no movements relatively to one another except the small movements of deformation which correspond to light-waves.

This theory—also called the theory of

the stationary luminiferous ether—moreover found a strong support in an experiment which is also of fundamental importance in the special theory of relativity, the experiment of Fizeau, from which one was obliged to infer that the luminiferous ether does not take part in the movements of bodies. The phenomenon of aberration also favoured the theory of the quasi-rigid ether.

The development of the theory of electricity along the path opened up by Maxwell and Lorentz gave the development of our ideas concerning the ether quite a peculiar and unexpected turn. For Maxwell himself the ether indeed still had properties which were purely mechanical, although of a much more complicated kind than the mechanical properties of tangible solid bodies. But neither Maxwell nor his followers succeeded in elaborating a mechanical model for the ether which might furnish a satisfactory mechanical interpretation of Maxwell's laws of the electro-magnetic field. The laws were clear and simple, the mechanical interpretations clumsy and contradictory. Almost imperceptibly the theoretical physicists adapted themselves to a situation which, from the standpoint of their mechanical programme, was very depressing. They were particularly influenced by the electro-dynamical investigations of Heinrich Hertz. For whereas they previously had required of a conclusive theory that it should content itself with the fundamental concepts which belong exclusively to mechanics (e.g. densities, velocities, deformations, stresses) they gradually accustomed themselves to admitting electric and magnetic force as fundamental concepts side by side with those of mechanics, without requiring a mechanical interpretation for them. Thus the purely mechanical view of nature was gradually abandoned. But this change led to a fundamental dualism which in the long-run was insupportable. A way of escape was now sought in the reverse direction, by reducing the principles of mechanics to those of electricity, and this especially as confidence in the strict validity of the equations of Newton's mechanics was shaken by the experiments with β-rays and rapid kathode rays.

This dualism still confronts us in unextenuated form in the theory of Hertz, where matter appears not only as the bearer of velocities, kinetic energy, and mechanical pressures, but also as the bearer of electromagnetic fields. Since such fields also occur *in vacuo*—i.e. in free ether—the ether also appears as bearer of electromagnetic fields. The ether appears indistinguishable in its functions from ordinary matter. Within matter it takes part in the motion of matter and in empty space it has everywhere a velocity; so that the ether has a definitely assigned velocity throughout the whole of space. There is no fundamental difference between Hertz's ether and ponderable matter (which in part subsists in the ether).

The Hertz theory suffered not only from the defect of ascribing to matter and ether, on the one hand mechanical states, and on the other hand electrical states, which do not stand in any conceivable relation to each other; it was also at variance with the result of Fizeau's important experiment on the velocity of the propagation of light in moving fluids, and with other established experimental results.

Such was the state of things when H. A. Lorentz entered upon the scene. He brought theory into harmony with experience by means of a wonderful simplification of theoretical principles. He achieved this, the most important advance in the theory of electricity since Maxwell, by taking from ether its mechanical, and from matter its electromagnetic qualities. As in empty space, so too in the interior of material bodies, the ether, and not matter viewed atomistically, was exclusively the seat of electromagnetic fields. According to Lorentz the elementary particles of matter alone are capable of carrying out movements; their electromagnetic activity is entirely confined to the carrying of electric charges. Thus Lorentz succeeded in reducing all electromagnetic happenings to Maxwell's equations for free space.

As to the mechanical nature of the

Lorentzian ether, it may be said of it, in a somewhat playful spirit, that immobility is the only mechanical property of which it has not been deprived by H. A. Lorentz. It may be added that the whole change in the conception of the ether which the special theory of relativity brought about, consisted in taking away from the ether its last mechanical quality, namely, its immobility. How this is to be understood will forthwith be expounded.

The space-time theory and the kinematics of the special theory of relativity were modelled on the Maxwell-Lorentz theory of the electromagnetic field. This theory therefore satisfies the conditions of the special theory of relativity, but when viewed from the latter it acquires a novel aspect. For if K be a system of co-ordinates relatively to which the Lorentzian ether is at rest, the Maxwell-Lorentz equations are valid primarily with reference to K. But by the special theory of relativity the same equations without any change of meaning also hold in relation to any new system of co-ordinates

K′ which is moving in uniform translation relatively to K. Now comes the anxious question:—Why must I in the theory distinguish the K system above all K′ systems, which are physically equivalent to it in all respects, by assuming that the ether is at rest relatively to the K system? For the theoretician such an asymmetry in the theoretical structure, with no corresponding asymmetry in the system of experience, is intolerable. If we assume the ether to be at rest relatively to K, but in motion relatively to K′, the physical equivalence of K and K′ seems to me from the logical standpoint, not indeed downright incorrect, but nevertheless inacceptable.

The next position which it was possible to take up in face of this state of things appeared to be the following. The ether does not exist at all. The electromagnetic fields are not states of a medium, and are not bound down to any bearer, but they are independent realities which are not reducible to anything else, exactly like the atoms of ponderable matter. This conception suggests itself the more readily as, according to Lorentz's theory, electromagnetic radiation, like ponderable matter, brings impulse and energy with it, and as, according to the special theory of relativity, both matter and radiation are but special forms of distributed energy, ponderable mass losing its isolation and appearing as a special form of energy.

More careful reflection teaches us, however, that the special theory of relativity does not compel us to deny ether. We may assume the existence of an ether; only we must give up ascribing a definite state of motion to it, i.e. we must by abstraction take from it the last mechanical characteristic which Lorentz had still left it. We shall see later that this point of view, the conceivability of which I shall at once endeavour to make more intelligible by a somewhat halting comparison, is justified by the results of the general theory of relativity.

Think of waves on the surface of water. Here we can describe two entirely different things. Either we may observe how the undulatory surface forming the boundary between water and air alters in the course of time; or else—with the help of small floats, for instance—we can observe how the position of the separate particles of water alters in the course of time. If the existence of such floats for tracking the motion of the particles of a fluid were a fundamental impossibility in physics—if, in fact, nothing else whatever were observable than the shape of the space occupied by the water as it varies in time, we should have no ground for the assumption that water consists of movable particles. But all the same we could characterise it as a medium.

We have something like this in the electromagnetic field. For we may picture the field to ourselves as consisting of lines of force. If we wish to interpret these lines of force to ourselves as something material in the ordinary sense, we are tempted to interpret the dynamic processes as motions of these lines of force, such that each separate line of force is tracked through the course of time. It is well known, however, that this way of regarding the electromagnetic field leads to contradictions.

Generalising we must say this :—There may be supposed to be extended physical objects to which the idea of motion cannot be applied. They may not be thought of as consisting of particles which allow themselves to be separately tracked through time. In Minkowski's idiom this is expressed as follows :—Not every extended conformation in the four-dimensional world can be regarded as composed of world-threads. The special theory of relativity forbids us to assume the ether to consist of particles observable through time, but the hypothesis of ether in itself is not in conflict with the special theory of relativity. Only we must be on our guard against ascribing a state of motion to the ether.

Certainly, from the standpoint of the special theory of relativity, the ether hypothesis appears at first to be an empty hypothesis. In the equations of the electromagnetic field there occur, in addition to the densities of the electric charge, *only* the intensities of the field. The career of electromagnetic processes *in vacuo* appears to be completely determined by these equations, uninfluenced by other physical quantities. The electromagnetic fields appear as ultimate, irreducible realities, and at first it seems superfluous to postulate a homogeneous, isotropic ether-medium, and to envisage electromagnetic fields as states of this medium.

But on the other hand there is a weighty argument to be adduced in favour of the ether hypothesis. To deny the ether is ultimately to assume that empty space has no physical qualities whatever. The fundamental facts of mechanics do not harmonize with this view. For the mechanical behaviour of a corporeal system hovering freely in empty space depends not only on relative positions (distances) and relative velocities, but also on its state of rotation, which physically may be taken as a characteristic not appertaining to the system in itself. In order to be able to look upon the rotation of the system, at least formally, as something real, Newton objectivises space. Since he classes his absolute space together with real things, for him rotation relative to an absolute space is also something real. Newton might no less well have called his absolute space "Ether"; what is essential is merely that besides observable objects, another thing, which is not perceptible, must be looked upon as real, to enable acceleration or rotation to be looked upon as something real.

It is true that Mach tried to avoid having to accept as real something which is not observable by endeavouring to substitute in mechanics a mean acceleration with reference to the totality of the masses in the universe in place of an acceleration with reference to absolute space. But inertial resistance opposed to relative acceleration of distant masses presupposes action at a distance; and as the modern physicist does not believe that he may accept this action at a distance, he comes back once more, if he follows Mach, to the ether, which has to serve as medium for the effects of inertia. But this conception of the ether to which we are led by Mach's way of thinking differs essentially from the ether as conceived by Newton, by Fresnel, and by Lorentz. Mach's ether not only *conditions* the behaviour of inert masses, but *is also conditioned* in its state by them.

Mach's idea finds its full development in the ether of the general theory of relativity. According to this theory the metrical qualities of the continuum of space-time differ in the environment of different points of space-time, and are partly conditioned by the matter existing outside of the territory under consideration. This space-time variability of the reciprocal relations of the standards of space and time, or, perhaps, the recognition of the fact that "empty space" in its physical relation is neither homogeneous nor isotropic, compelling us to describe its state by ten functions (the gravitation potentials $g_{\mu\nu}$), has, I think, finally disposed of the view that space is physically empty. But therewith the conception of the ether has again acquired an intelligible content, although this content differs widely from that of the ether of the mechanical undulatory theory of light. The ether of the general theory of relativity is a medium which is itself devoid of *all* mechanical and kinematical qualities, but helps to determine mechanical (and electromagnetic) events.

What is fundamentally new in the ether of the general theory of relativity as opposed to the ether of Lorentz consists in this, that the state of the former is at every place determined by connections with the matter and the state of the ether in neighbouring places, which are amenable to law in the form of differential equations; whereas the state of the Lorentzian ether in the absence of electromagnetic fields is conditioned by nothing outside itself, and is everywhere the same. The ether of the general theory of relativity is transmuted conceptually into the ether of Lorentz if we substitute constants for the functions of space which describe the former, disregarding the causes which condition its state. Thus we may also say, I think, that the ether of the general theory of relativity is the outcome of the Lorentzian ether, through relativation.

As to the part which the new ether is to play in the physics of the future we are not yet clear. We know that it determines the metrical relations in the space-time continuum, e.g. the configurative possibilities of solid bodies as well as the gravitational fields ; but we do not know whether it has an essential share in the structure of the electrical elementary particles constituting matter. Nor do we know whether it is only in the proximity of ponderable masses that its structure differs essentially from that of the Lorentzian ether ; whether the geometry of spaces of cosmic extent is approximately Euclidean. But we can assert by reason of the relativistic equations of gravitation that there must be a departure from Euclidean relations, with spaces of cosmic order of magnitude, if there exists a positive mean density, no matter how small, of the matter in the universe.

In this case the universe must of necessity be spatially unbounded and of finite magnitude, its magnitude being determined by the value of that mean density.

If we consider the gravitational field and the electromagnetic field from the standpoint of the ether hypothesis, we find a remarkable difference between the two. There can be no space nor any part of space without gravitational potentials; for these confer upon space its metrical qualities, without which it cannot be imagined at all. The existence of the gravitational field is inseparably bound up with the existence of space. On the other hand a part of space may very well be imagined without an electromagnetic field; thus in contrast with the gravitational field, the electromagnetic field seems to be only secondarily linked to the ether, the formal nature of the electromagnetic field being as yet in no way determined by that of gravitational ether. From the present state of theory it looks as if the electromagnetic field, as opposed to the gravitational field, rests upon an entirely new formal *motif*, as though nature might just as well have endowed the gravitational ether with fields of quite another type, for example, with fields of a scalar potential, instead of fields of the electromagnetic type.

Since according to our present conceptions the elementary particles of matter are also, in their essence, nothing else than condensations of the electromagnetic field, our present view of the universe presents two realities which are completely separated from each other conceptually, although connected causally, namely, gravitational ether and electromagnetic field, or—as they might also be called—space and matter.

Of course it would be a great advance if we could succeed in comprehending the gravitational field and the electromagnetic field together as one unified conformation. Then for the first time the epoch of theoretical physics founded by Faraday and Maxwell would reach a satisfactory conclusion. The contrast between ether and matter would fade away, and, through the general theory of relativity, the whole of physics would become a complete system of thought, like geometry, kinematics, and the theory of gravitation. An exceedingly ingenious attempt in this direction has been made by the mathematician H. Weyl; but I do not believe that his theory will hold its ground in relation to reality. Further, in contemplating the immediate future of theoretical physics we ought not unconditionally to reject the possibility that the facts comprised in the quantum theory may set bounds to the field theory beyond which it cannot pass.

Recapitulating, we may say that according to the general theory of relativity space is endowed with physical qualities; in this sense, therefore, there exists an ether. According to the general theory of relativity space without ether is unthinkable; for in such space there not only would be no propagation of light, but also no possibility of existence for standards of space and time (measuring-rods and clocks), nor therefore any space-time intervals in the physical sense. But this ether may not be thought of as endowed with the quality characteristic of ponderable media, as consisting of parts which may be tracked through time. The idea of motion may not be applied to it.

GEOMETRY AND EXPERIENCE

An expanded form of an Address to the Prussian Academy of Sciences in Berlin on January 27th, 1921.

GEOMETRY AND EXPERIENCE

ONE reason why mathematics enjoys special esteem, above all other sciences, is that its laws are absolutely certain and indisputable, while those of all other sciences are to some extent debatable and in constant danger of being overthrown by newly discovered facts. In spite of this, the investigator in another department of science would not need to envy the mathematician if the laws of mathematics referred to objects of our mere imagination, and not to objects of reality. For it cannot occasion surprise that different persons should arrive at the same logical conclusions when they have already agreed upon the fundamental laws (axioms), as well as the methods by which other laws are to be deduced therefrom. But there is another reason for the high repute of mathematics, in that it is mathematics which affords the exact natural sciences a certain measure of security, to which without mathematics they could not attain.

At this point an enigma presents itself which in all ages has agitated inquiring minds. How can it be that mathematics, being after all a product of human thought which is independent of experience, is so admirably appropriate to the objects of reality? Is human reason, then, without experience, merely by taking thought, able to fathom the properties of real things.

In my opinion the answer to this question is, briefly, this:—As far as the laws of mathematics refer to reality, they are not certain; and as far as they are certain, they do not refer to reality. It seems to me that complete clearness as to this state of things first became common property through that new departure in mathematics which is known by the name of mathematical logic or "Axiomatics." The progress achieved by axiomatics consists in its having neatly separated the logical-formal from its objective or intuitive content; according to axiomatics the logical-formal alone forms the subject-matter of mathematics, which is not concerned with the intuitive or other content associated with the logical-formal.

Let us for a moment consider from this point of view any axiom of geometry, for instance, the following:—Through two points in space there always passes one and only one straight line. How is this axiom to be interpreted in the older sense and in the more modern sense?

The older interpretation:—Every one knows what a straight line is, and what a point is. Whether this knowledge springs from an ability of the human mind or from experience, from some collaboration of the two or from some other source, is not for the mathematician to decide. He leaves the question to the philosopher. Being based upon this knowledge, which precedes all mathematics, the axiom stated above is, like all other axioms, self-evident, that is, it is the expression of a part of this *à priori* knowledge.

The more modern interpretation:—Geometry treats of entities which are denoted by the words straight line, point, etc. These entities do not take for granted any knowledge or intuition whatever, but they presuppose only the validity of the axioms, such as the one stated above, which are to be taken in a purely formal sense, i.e. as void of all content of intuition or experience. These axioms are free creations of the human mind. All other propositions of geometry are logical inferences from the axioms (which are to be taken in the nominalistic sense only). The matter of which geometry treats is first defined by the axioms. Schlick in his book on epistemology has therefore characterised axioms very aptly as "implicit definitions."

This view of axioms, advocated by modern axiomatics, purges mathematics of all extraneous elements, and thus dispels the mystic obscurity which formerly surrounded the principles of mathematics. But a presentation of its principles thus clarified makes it also evident that mathematics as such cannot predicate anything about perceptual objects or real objects. In axiomatic geometry the words "point," "straight line," etc., stand only for empty conceptual schemata. That which gives them substance is not relevant to mathematics.

Yet on the other hand it is certain that mathematics generally, and particularly geometry, owes its existence to the need which was felt of learning something about the relations of real things to one another. The very word geometry, which, of course, means earth-measuring, proves this. For earth-measuring has to do with the possibilities of the disposition of certain natural objects with respect to one another, namely, with parts of the earth, measuring-lines, measuring-wands, etc. It is clear that the system of concepts of axiomatic geometry alone cannot make any assertions as to the relations of real objects of this kind, which we will call practically-rigid bodies. To be able to make such assertions, geometry must be stripped of its merely logical-formal character by the co-ordination of real objects of experience with the empty conceptual frame-work of axiomatic geometry. To accomplish this, we need only add the proposition :—Solid bodies are related, with respect to their possible dispositions, as are bodies in Euclidean geometry of three dimensions. Then the propositions of Euclid contain affirmations as to the relations of practically-rigid bodies.

Geometry thus completed is evidently a natural science ; we may in fact regard it as the most ancient branch of physics. Its affirmations rest essentially on induction from experience, but not on logical inferences only. We will call this completed geometry "practical geometry," and shall distinguish it in what follows from "purely axiomatic geometry." The question whether the practical geometry of the universe is Euclidean or not has a clear meaning, and its answer can only be furnished by experience. All linear measurement in physics is practical geometry in this sense, so too is geodetic and astronomical linear measurement, if we call to our help the law of experience that light is propagated in a straight line, and indeed in a straight line in the sense of practical geometry.

I attach special importance to the view of geometry which I have just set forth, because without it I should have been unable to formulate the theory of relativity. Without it the following reflection would have been impossible :—In a system of reference rotating relatively to an inert system, the laws of disposition of rigid bodies do not correspond to the rules of Euclidean geometry on account of the Lorentz contraction ; thus if we admit non-inert systems we must abandon Euclidean geometry. The decisive step in the transition to general co-variant equations would certainly not have been taken if the above interpretation had not served as a stepping-stone. If we deny the relation between the body of axiomatic Euclidean geometry and the practically-rigid body of reality, we readily arrive at the following view, which was entertained by that acute and profound thinker,

H. Poincaré :—Euclidean geometry is distinguished above all other imaginable axiomatic geometries by its simplicity. Now since axiomatic geometry by itself contains no assertions as to the reality which can be experienced, but can do so only in combination with physical laws, it should be possible and reasonable—whatever may be the nature of reality—to retain Euclidean geometry. For if contradictions between theory and experience manifest themselves, we should rather decide to change physical laws than to change axiomatic Euclidean geometry. If we deny the relation between the practically-rigid body and geometry, we shall indeed not easily free ourselves from the convention that Euclidean geometry is to be retained as the simplest. Why is the equivalence of the practically-rigid body and the body of geometry—which suggests itself so readily—denied by Poincaré and other investigators? Simply because under closer inspection the real solid bodies in nature are not rigid, because their geometrical behaviour, that is, their possibilities of relative disposition, depend upon temperature, external forces, etc. Thus the original, immediate relation between geometry and physical reality appears destroyed, and we feel impelled toward the following more general view, which characterizes Poincaré's standpoint. Geometry (G) predicates nothing about the relations of real things, but only geometry together with the purport (P) of physical laws can do so. Using symbols, we may say that only the sum of (G) + (P) is subject to the control of experience. Thus (G) may be chosen arbitrarily, and also parts of (P); all these laws are conventions. All that is necessary to avoid contradictions is to choose the remainder of (P) so that (G) and the whole of (P) are together in accord with experience. Envisaged in this way, axiomatic geometry and the part of natural law which has been given a conventional status appear as epistemologically equivalent.

Sub specie aeterni Poincaré, in my opinion, is right. The idea of the measuring-rod and the idea of the clock co-ordinated with it in the theory of relativity do not find their exact correspondence in the real world. It is also clear that the solid body and the clock do not in the conceptual edifice of physics play the part of irreducible elements, but that of composite structures, which may not play any independent part in theoretical physics. But it is my conviction that in the present stage of development of theoretical physics these ideas must still be employed as independent ideas; for we are still far from possessing such certain knowledge of theoretical principles as to be able to give exact theoretical constructions of solid bodies and clocks.

Further, as to the objection that there are no really rigid bodies in nature, and that therefore the properties predicated of rigid bodies do not apply to physical reality,—this objection is by no means so radical as might appear from a hasty examination. For it is not a difficult task to determine the physical state of a measuring-rod so accurately that its behaviour relatively to other measuring-bodies shall be sufficiently free from ambiguity to allow it to be substituted for the "rigid" body. It is to measuring-bodies of this kind that statements as to rigid bodies must be referred.

All practical geometry is based upon a principle which is accessible to experience, and which we will now try to realise. We will call that which is enclosed between two boundaries, marked upon a practically-rigid body, a tract. We imagine two practically-rigid bodies, each with a tract marked out on it. These two tracts are said to be "equal to one another" if the boundaries of the one tract can be brought to coincide permanently with the boundaries of the other. We now assume that:

If two tracts are found to be equal once and anywhere, they are equal always and everywhere.

Not only the practical geometry of Euclid, but also its nearest generalisation, the practical geometry of Riemann, and therewith the general theory of relativity, rest upon this assumption. Of the experimental reasons which warrant this assumption I will mention only one. The phenomenon of the propagation of light in empty space assigns a tract, namely, the appropriate path of light, to each interval of local time, and conversely. Thence it follows that the above assumption for tracts must also hold good for intervals of clock-time in the theory of relativity. Consequently it may be formulated as follows:—If two ideal clocks are going at the same rate at any time and at any place (being then in immediate proximity to each other), they will always go at the same rate, no matter where and when they are again compared with each other at one place.—If this law were not valid for real clocks, the proper frequencies for the separate atoms of the same chemical element would not be in such exact agreement as experience demonstrates. The existence of sharp spectral lines is a convincing experimental proof of the above-mentioned principle of practical geometry. This is the ultimate foundation in fact which enables us to speak with meaning of the mensuration, in Riemann's sense of the word, of the four-dimensional continuum of space-time.

The question whether the structure of this continuum is Euclidean, or in accordance with Riemann's general scheme, or otherwise, is, according to the view which is here being advocated, properly speaking a physical question which must be answered by experience, and not a question of a mere convention to be selected on practical grounds. Riemann's geometry will be the right thing if the laws of disposition of practically-rigid bodies are transformable into those of the bodies of Euclid's geometry with an exactitude which increases in proportion as the dimensions of the part of space-time under consideration are diminished.

It is true that this proposed physical interpretation of geometry breaks down when applied immediately to spaces of sub-molecular order of magnitude. But nevertheless, even in questions as to the constitution of elementary particles, it retains part of its importance. For even when it is a question of describing the electrical elementary particles constituting matter, the attempt may still be made to ascribe physical importance to those ideas of fields which have been physically defined for the purpose of describing the geometrical behaviour of bodies which are large as compared with the molecule. Success alone can decide as to the justification of such an attempt, which postulates physical reality for the fundamental principles of Riemann's geometry outside of the domain of their physical definitions. It might possibly turn out that this extrapolation has no better warrant than the extrapolation of the idea of temperature to parts of a body of molecular order of magnitude.

It appears less problematical to extend the ideas of practical geometry to spaces of cosmic order of magnitude. It might, of course, be objected that a construction composed of solid rods departs more and more from ideal rigidity in proportion as its spatial extent becomes greater.

But it will hardly be possible, I think, to assign fundamental significance to this objection. Therefore the question whether the universe is spatially finite or not seems to me decidedly a pregnant question in the sense of practical geometry. I do not even consider it impossible that this question will be answered before long by astronomy. Let us call to mind what the general theory of relativity teaches in this respect. It offers two possibilities :—

1. The universe is spatially infinite. This can be so only if the average spatial density of the matter in universal space, concentrated in the stars, vanishes, i.e. if the ratio of the total mass of the stars to the magnitude of the spacc through which they are scattered approximates indefinitely to the value zero when the spaces taken into consideration are constantly greater and greater.

2. The universe is spatially finite. This must be so, if there is a mean density of the ponderable matter in universal space differing from zero. The smaller

that mean density, the greater is the volume of universal space.

I must not fail to mention that a theoretical argument can be adduced in favour of the hypothesis of a finite universe. The general theory of relativity teaches that the inertia of a given body is greater as there are more ponderable masses in proximity to it; thus it seems very natural to reduce the total effect of inertia of a body to action and reaction between it and the other bodies in the universe, as indeed, ever since Newton's time, gravity has been completely reduced to action and reaction between bodies. From the equations of the general theory of relativity it can be deduced that this total reduction of inertia to reciprocal action between masses—as required by E. Mach, for example—is possible only if the universe is spatially finite.

On many physicists and astronomers this argument makes no impression. Experience alone can finally decide which of the two possibilities is realised in nature. How can experience furnish an answer? At first it might seem possible to determine the mean density of matter by observation of that part of the universe which is accessible to our perception. This hope is illusory. The distribution of the visible stars is extremely irregular, so that we on no account may venture to set down the mean density of star-matter in the universe as equal, let us say, to the mean density in the Milky Way. In any case, however great the space examined may be, we could not feel convinced that there were no more stars beyond that space. So it seems impossible to estimate the mean density.

But there is another road, which seems to me more practicable, although it also presents great difficulties. For if we inquire into the deviations shown by the consequences of the general theory of relativity which are accessible to experience, when these are compared with the consequences of the Newtonian theory, we first of all find a deviation which shows itself in close proximity to gravitating mass, and has been confirmed in the case of the planet

Mercury. But if the universe is spatially finite there is a second deviation from the Newtonian theory, which, in the language of the Newtonian theory, may be expressed thus :—The gravitational field is in its nature such as if it were produced, not only by the ponderable masses, but also by a mass-density of negative sign, distributed uniformly throughout space. Since this factitious mass-density would have to be enormously small, it could make its presence felt only in gravitating systems of very great extent.

Assuming that we know, let us say, the statistical distribution of the stars in the Milky Way, as well as their masses, then by Newton's law we can calculate the gravitational field and the mean velocities which the stars must have, so that the Milky Way should not collapse under the mutual attraction of its stars, but should maintain its actual extent. Now if the actual velocities of the stars, which can, of course, be measured, were smaller than the calculated velocities, we should have a proof that the actual attractions at great distances are smaller than by Newton's law. From such a deviation it could be proved indirectly that the universe is finite. It would even be possible to estimate its spatial magnitude.

Can we picture to ourselves a three-dimensional universe which is finite, yet unbounded?

The usual answer to this question is "No," but that is not the right answer. The purpose of the following remarks is to show that the answer should be "Yes." I want to show that without any extraordinary difficulty we can illustrate the theory of a finite universe by means of a mental image to which, with some practice, we shall soon grow accustomed.

First of all, an obervation of epistemological nature. A geometrical-physical theory as such is incapable of being directly pictured, being merely a system of concepts. But these concepts serve the purpose of bringing a multiplicity of real or imaginary sensory experiences into connection in the mind. To "visualise" a theory, or bring it home to one's mind, therefore means to give a representation to that abundance of experiences for which the theory supplies the schematic arrangement. In the present case we have to ask ourselves how we can represent that relation of solid bodies with respect to their reciprocal disposition (contact) which corresponds to the theory of a finite universe. There is really nothing new in what I have to say about this; but innumerable questions addressed to me prove that the requirements of those who thirst for knowledge of these matters have not yet been completely satisfied. So, will the initiated please pardon me, if part of what I shall bring forward has long been known?

What do we wish to express when we say that our space is infinite? Nothing more than that we might lay any number whatever of bodies of equal sizes side by side without ever filling space. Suppose that we are provided with a great many wooden cubes all of the same size. In accordance with Euclidean geometry we can place them above, beside, and behind one another so as to fill a part of space of any dimensions; but this construction would never be finished; we could go on adding more and more cubes without ever finding that there was no more room. That is what we wish to express when we say that space is infinite. It would be better to say that space is infinite in relation to practically-rigid bodies, assuming that the laws of disposition for these bodies are given by Euclidean geometry.

Another example of an infinite continuum is the plane. On a plane surface we may lay squares of cardboard so that each side of any square has the side of another square adjacent to it. The construction is never finished; we can always go on laying squares—if their laws of disposition correspond to those of plane figures of Euclidean geometry. The plane is therefore infinite in relation to the cardboard squares. Accordingly we say that the plane is an infinite continuum of two dimensions, and space an infinite continuum of three dimensions. What is here meant by the number of dimensions, I think I may assume to be known.

Now we take an example of a two-dimensional continuum which is finite, but unbounded. We imagine the surface of a large globe and a quantity of small paper discs, all of the same size. We place one of the discs anywhere on the surface of the globe. If we move the disc about, anywhere we like, on the surface of the globe, we do not come upon a limit or boundary anywhere on the journey. Therefore we say that the spherical surface of the globe is an unbounded continuum. Moreover, the spherical surface is a finite continuum. For if we stick the paper discs on the globe, so that no disc overlaps another, the surface of the globe will finally become so full that there is no room for another disc. This simply means that the spherical surface of the globe is finite in relation to the paper discs. Further, the spherical surface is a non-Euclidean continuum of two dimensions, that is to say, the laws of disposition for the rigid figures lying in it do not agree with those of the Euclidean plane. This can be shown in the following way.

Place a paper disc on the spherical surface, and around it in a circle place six more discs, each of which is to be surrounded in turn by six discs, and so on. If this construction is made on a plane surface, we have an uninterrupted disposition in which there are six discs touching every disc except those which lie on the outside.

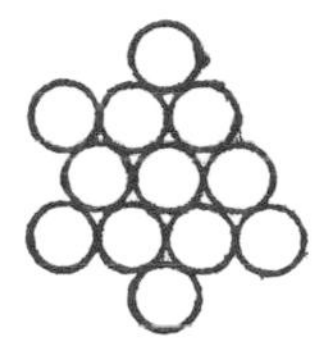

FIG. 1.

On the spherical surface the construction also seems to promise success at the outset, and the smaller the radius of the disc in proportion to that of the sphere, the more promising it seems. But as the construction progresses it becomes more and more patent that the disposition of the discs in the manner indicated, without interruption, is not possible, as it should be possible by Euclidean geometry of the the plane surface. In this way creatures which cannot leave the spherical surface, and cannot even peep out from the spherical surface into three-dimensional space, might discover, merely by experimenting with discs, that their two-dimensional " space " is not Euclidean, but spherical space.

From the latest results of the theory of relativity it is probable that our three-dimensional space is also approximately spherical, that is, that the laws of disposition of rigid bodies in it are not given by Euclidean geometry, but approximately by spherical geometry, if only we consider parts of space which are sufficiently great. Now this is the place where the reader's imagination boggles. " Nobody can imagine this thing," he cries indignantly. " It can be said, but cannot be thought. I can represent to myself a spherical surface well enough, but nothing analogous to it in three dimensions."

We must try to surmount this barrier in the mind, and the patient reader will see that it is by no means a particularly difficult task. For this purpose we will first give our attention once more to the geometry of two-dimensional spherical surfaces. In the adjoining figure let K be the spherical surface, touched at S by a plane, E, which, for facility of presentation, is shown in the drawing as a bounded surface. Let L be a disc on the spherical surface. Now let us imagine that at the point N of the spherical surface,

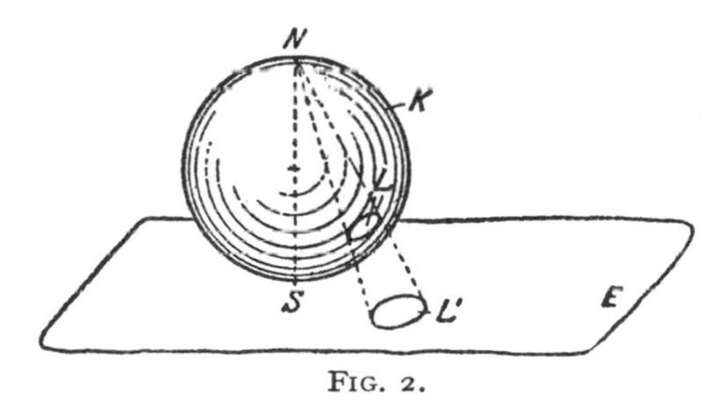

FIG. 2.

diametrically opposite to S, there is a luminous point, throwing a shadow L' of the disc L upon the plane E. Every point on the sphere has its shadow on the plane. If the disc on the sphere K is moved, its shadow L' on the plane E also moves. When the disc L is at S, it almost exactly coincides with its shadow. If it moves on the spherical surface away from S upwards, the disc shadow L' on the plane also moves away from S on the plane outwards, growing bigger and bigger. As the disc L approaches the luminous point N, the shadow moves off to infinity, and becomes infinitely great.

Now we put the question, What are the laws of disposition of the disc-shadows L' on the plane E? Evidently they are exactly the same as the laws of disposition of the discs L on the spherical surface. For to each original figure on K there is a corresponding shadow figure on E. If two discs on K are touching, their shadows on E also touch. The shadow-geometry on the plane agrees with the the disc-geometry on the sphere. If we call the disc-shadows rigid figures, then spherical geometry holds good on the plane E with respect to these rigid figures. Moreover, the plane is finite with respect to the disc-shadows, since only a finite number of the shadows can find room on the plane.

At this point somebody will say, "That is nonsense. The disc-shadows are *not* rigid figures. We have only to move a two-foot rule about on the plane E to convince ourselves that the shadows constantly increase in size as they move away from S on the plane towards infinity." But what if the two-foot rule were to behave on the plane E in the same way as the disc-shadows L'? It would then be impossible to show that the shadows increase in size as they move away from S; such an assertion would then no longer have any meaning whatever. In fact the only objective assertion that can be made about the disc-shadows is just this, that they are related in exactly the same way as are the rigid discs on the spherical surface in the sense of Euclidean geometry.

We must carefully bear in mind that our statement as to the growth of the disc-shadows, as they move away from S towards infinity, has in itself no objective meaning, as long as we are unable to employ Euclidean rigid bodies which can be moved about on the plane E for the purpose of comparing the size of the disc-shadows. In respect of the laws of disposition of the shadows L', the point S has no special privileges on the plane any more than on the spherical surface.

The representation given above of spherical geometry on the plane is important for us, because it readily allows itself to be transferred to the three-dimensional case.

Let us imagine a point S of our space, and a great number of small spheres, L', which can all be brought to coincide with one another. But these spheres are not to be rigid in the sense of Euclidean geometry; their radius is to increase (in the sense of Euclidean geometry) when they are moved away from S towards infinity, and this increase is to take place in exact accordance with the same law as applies to the increase of the radii of the disc-shadows L′ on the plane.

After having gained a vivid mental image of the geometrical behaviour of our L' spheres, let us assume that in our space there are no rigid bodies at all in the sense of Euclidean geometry, but only bodies having the behaviour of our L' spheres.

Then we shall have a vivid representation of three-dimensional spherical space, or, rather of three-dimensional spherical geometry. Here our spheres must be called "rigid" spheres. Their increase in size as they depart from S is not to be detected by measuring with measuring-rods, any more than in the case of the disc-shadows on E, because the standards of measurement behave in the same way as the spheres. Space is homogeneous, that is to say, the same spherical configurations are possible in the environment of all points.[1] Our space is finite, because, in consequence of the "growth" of the spheres, only a finite number of them can find room in space.

In this way, by using as stepping-stones the practice in thinking and visualisation which Euclidean geometry gives us, we have acquired a mental picture of spherical geometry. We may without difficulty

[1] This is intelligible without calculation—but only for the two-dimensional case—if we revert once more to the case of the disc on the surface of the sphere.

impart more depth and vigour to these ideas by carrying out special imaginary constructions. Nor would it be difficult to represent the case of what is called elliptical geometry in an analogous manner. My only aim to-day has been to show that the human faculty of visualisation is by no means bound to capitulate to non-Euclidean geometry.